T0236880

SpringerBriefs in Applied Sciences and Technology

Computational Intelligence

Series editor

Janusz Kacprzyk, Polish Academy of Sciences, Systems Research Institute, Warsaw, Poland

The series "Studies in Computational Intelligence" (SCI) publishes new developments and advances in the various areas of computational intelligence—quickly and with a high quality. The intent is to cover the theory, applications, and design methods of computational intelligence, as embedded in the fields of engineering, computer science, physics and life sciences, as well as the methodologies behind them. The series contains monographs, lecture notes and edited volumes in computational intelligence spanning the areas of neural networks, connectionist systems, genetic algorithms, evolutionary computation, artificial intelligence, cellular automata, self-organizing systems, soft computing, fuzzy systems, and hybrid intelligent systems. Of particular value to both the contributors and the readership are the short publication timeframe and the world-wide distribution, which enable both wide and rapid dissemination of research output.

More information about this series at http://www.springer.com/series/10618

Ali Mohammad Saghiri ·
M. Daliri Khomami · Mohammad Reza Meybodi

Intelligent Random Walk: An Approach Based on Learning Automata

 Springer

Ali Mohammad Saghiri
Amirkabir University of Technology
Tehran, Iran

M. Daliri Khomami
Amirkabir University of Technology
Tehran, Iran

Mohammad Reza Meybodi
Amirkabir University of Technology
Tehran, Iran

ISSN 2191-530X   ISSN 2191-5318   (electronic)
SpringerBriefs in Applied Sciences and Technology
ISSN 2625-3704   ISSN 2625-3712   (electronic)
SpringerBriefs in Computational Intelligence
ISBN 978-3-030-10882-3   ISBN 978-3-030-10883-0   (eBook)
https://doi.org/10.1007/978-3-030-10883-0

Library of Congress Control Number: 2018966119

This Springer imprint is published by the registered company Springer Nature Switzerland AG
The registered company address is: Gewerbestrasse 11, 6330 Cham, Switzerland

Preface

Recently, random walk algorithms have attracted considerable attention because they are easy to interpret. In addition, they are used to problem-solving, modeling, and simulation in many types of networks including computer networks, social networks, and biological networks. Recently, the k-random walk algorithm is reported to improve the capabilities of the primary version of the random walk algorithm. This type of random walk algorithm is a faster version of the primary version of the random walk algorithm. When the random walk algorithms are used to solve the problems in real-world applications, several challenges will be raised. In real-world applications, both k-random walk and random walk algorithms must be tuned considering information about the nature of the application. Considering the mentioned issue, intelligent models of random walk have been reported in the literature. These models of random walk may be used to solve a wide range of problems in real-world applications. In recent years, several intelligent models of random walk are reported based on learning automata. Learning automata are a type of machine learning algorithms called reinforcement learning. In this book, we study the intelligent models of random walk based on learning automata. These models of random walk try to gradually learn required information from the nature of the application to improve their efficiency. We also study the corresponding applications of these models of random walk. More precisely, we studied the applicability of the intelligent models of random walk as efficient prediction models for two large-scale networks such as peer-to-peer networks and social networks. This book opens a new horizon for designing prediction models and problem-solving methods based on intelligent models of random walk. The challenges and open problems raised by this approach will be also studied in this book.

Tehran, Iran

Ali Mohammad Saghiri
M. Daliri Khomami
Mohammad Reza Meybodi

Contents

Acronyms and Abbreviations

AI	Artificial Intelligence
CA	Cellular Automaton
CALA	Continuous Action Learning Automata
FALA	Finite Action Learning Automata
FSLA	Fixed Structure Learning Automata
GLA	Generalized Learning Automata
IKRW-LA	Intelligent K-Random Walk based on Learning Automata
IRW-LA	Intelligent Random Walk based on Learning Automata
LA	Learning Automaton
ML	Machine Learning
NN	Neural Network
P2P	Peer-to-Peer
PLA	Parameterized Learning Automata
RL	Reinforcement Learning
RW	Random Walk
SOIRW-LA	Self-Organized Intelligent Random Walk based on Learning Automata
VSLA	Variable Structure Learning Automata

Chapter 1
Random Walk Algorithms: Definitions, Weaknesses, and Learning Automata-Based Approach

Abstract Random walk algorithms are used to problem-solving, modeling, and simulation in many types of networks including computer networks, social networks, and biological networks. In real-world problems, the non-intelligent models of random walk may not be used as a problem-solving method. Recently, intelligent models of random walk have been reported in the literature. These models try to extend the basic versions of random walk to design a novel problem-solving method. The learning mechanism of these models is based on learning automata. In these models, the design of feedback systems given by the theory of learning automata is used to design intelligent models of random walk. In this chapter, we discuss about the weaknesses of non-intelligent models of random walk as a problem-solving method in real-world applications. We also give the required information about random walk algorithms and the theory of learning automata.

Keywords Random walk algorithm · Networks · Feedback systems · Learning automata

1.1 Introduction

A random walk is one of the simplest dynamical processes that can occur on a network. Random walk algorithms on networks have attracted considerable interest because they are applicable for problem-solving and also easy to interpret. The random walk algorithms also can be used to approximate many natural processes. Different types of random walk algorithms are reported in the literature. They have yielded important insights into a huge variety of applications (Fig. 1.1). Some of these applications are explained as below.

- **Web computing**: Random walk algorithms have been applied to rank web pages and sports teams, optimize searches, investigate the efficiency of network navigation, characterize cyclic structures in networks, and coarse-grain networks to highlight mesoscale features such as community structure [1, 2].

A. M. Saghiri et al., *Intelligent Random Walk: An Approach Based on Learning Automata*, SpringerBriefs in Computational Intelligence, https://doi.org/10.1007/978-3-030-10883-0_1

Fig. 1.1 Applications of
random walk algorithms

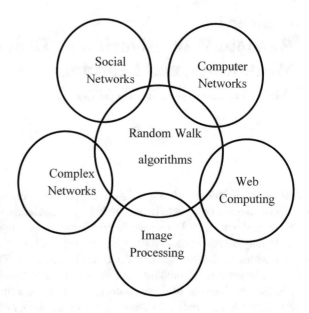

- **Complex networks**: Another interesting application of random walk algorithm is to calculate the centrality of actors in complex networks when there is no knowledge about the full network topology but only local information is available [3–5].
- **Computer networks**: In computer networks such as peer-to-peer networks, the random walk algorithms are used to design efficient search algorithms. Since the scale of these networks is very large, designing an appropriate search algorithm for them is a challenging problem. Many algorithms such as those reported in [6–10] utilize intelligent models of random walk.
- **Social networks**: In the context of complex social networks, random walks have proven to be useful tools and several algorithms have been proposed for structural properties of the networks [11–13]. Among them, the restricted dynamics of Self-Avoiding Random Walks (SAW) [6] in the application of community detection is proposed. Based on this algorithm, random walker visits only at most once each vertex [14].
- **Image Processing**: Random walk algorithms are applied in image segmentation to determine the label of objects which associate with each pixel. With the aid of random walk, a small number of pixels with user-defined (or predefined) labels, one can analytically and quickly determine the probability that a random walker starting at each unlabeled pixel will first reach to one of the pre-labeled pixels. This algorithm is typically referred to as the random walker segmentation algorithm [15].

In the other hand, a learning automaton is an adaptive decision-making unit in which the performance is improved by learning how to choose the optimal action from a finite set of allowed actions considering repeated interactions with a random environment. Learning automata are a type of machine learning algorithm called reinforcement learning. Recently, different versions of intelligent random walk algorithms based on learning automata are reported in the literature. In this book, we summarize the recent approaches for implementing an intelligent random walk based on learning automata.

1.2 Basic Concepts

In this section, in order to provide basic information for the remainder of this book, we present a brief overview of random walk algorithms, k-random walk algorithms, and theory of learning automata.

1.2.1 Random Walk Algorithms

Random walk algorithms have attracted considerable attention because they are easy to interpret. In the following, we will describe the behavior of random walk algorithms in the networks. In a random walk on a graph, the graph and a starting node are given. During a walk on the graph, we select a neighbor of the node at random manner. After selecting the neighbor, we move to the neighbor. Then, we select a neighbor of the node at random, and move to it. During this procedure, a sequence of nodes is constructed which determine a traverse for the graph [16–18].

1.2.2 K-Random Walk Algorithms

It is obvious that executing multiple random walk algorithms on a graph result in faster than a single random walk in searching a network. In k-random walk algorithms, we choose a random node in the network and trigger k-random neighbors of that node. All of the triggered nodes repeat this process. Recently, these types of algorithms are used to search in large-scale networks such as peer-to-peer networks and social networks [6–10]. This is because random walk algorithms have very simple logic and faster version of these algorithms may be applied in new generations of large-scale networks reported in Internet of Things (IoT), complex networks, and grid computing.

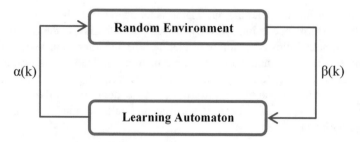

Fig. 1.2   A learning automaton and the relationship with its random environ

1.2.3   Theory of Learning Automata

Learning automata are a type of machine learning algorithms called reinforcement learning. A Learning Automaton (LA) is an adaptive decision-making unit in which the performance is improved by learning how to choose the optimal action from a finite set of allowed actions considering repeated interactions with a random environment [19]. The action is chosen by the LA. In turn, the environment responds to the action taken with a reinforcement signal. The objective of an LA is to find the optimal action from the action set so that the average reward received from the environment is maximized. The relationship between the LA and its random environment is shown in Fig. 1.2. Several learning automata-based algorithms have been proposed in the application of social networks [20–22] and peer-to-peer networks [23–25].

1.3   Random Walk Weaknesses and the Approach of Learning Automata

Random walk algorithms are used to problem-solving, modeling, and simulation in many types of applications. In real-world applications, the random walk algorithms must be tuned considering information about the nature of the application. The performance of non-intelligent models of random walk is low in practical problems because these models do not consider the changes and information about the nature of the practical problems. To solve this problem, in real-world applications, we may use feedback loops to improve the performance of random walk algorithms. Recently, the intelligent models of the random walk are reported in the literature [6, 7]. The rationale behind intelligent models based on random walk algorithms and learning automata is to extend the capabilities of random walk algorithms by the feedback loops of the theory of learning automata.

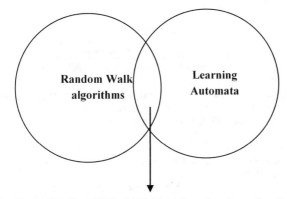

Intelligent Random Walk Algorithms based on Learning Automata

Fig. 1.3 A conceptual diagram for combination of random walk and learning automata

Figure 1.3 shows a conceptual diagram for combination of random walk algorithms and theory of learning automata. In the rest of this section, the main problems of the random walk algorithms in the networks for real-world applications are explained.

- The first problem is to find the optimum value of the k in k-random walk algorithm in every arbitrary network. Choosing a value of k larger than the average number of neighbors of all nodes of the network results in observing useless parts of the network. For example, in a computer network, a large value for k results in generating more traffic in the computer network, because it works like flooding search methods. On the other hand, if k is smaller than the average number of neighbors of all nodes of the network, the probability of selecting desirable parts of the network is decreased. For example, in a computer network, a low value for k results in generating low traffic and limited observation which leads to miss the appropriate nodes.
- The second problem is to determine which neighbors should be selected in each node of the network.

Both above mentioned problems can be solved using feedback system of the theory of learning automata. Intelligent models of random walk based on learning automata may be used to a wide range of real-world applications. As it was previously mentioned, multiple intelligent models of random walk based on learning automata are reported in [6, 7]. These models of random walk try to gradually learn the required information from the nature of the application to improve their efficiency. In the rest of this book, we suggest three intelligent models of the random walk based on learning automata. The proposed models will be implicitly used as prediction models for problem-solving in large-scale complex networks such as peer-to-peer networks and social networks.

References

1. Bar-Yossef Z, Berg A, Chien S, Fakcharoenphol J, Weitz D (2000) Approximating aggregate queries about web pages via random walks. In: Proceedings of the 26th international conference on very large data bases. ACM, Egypt, pp 535–544
2. Tong H, Faloutsos C, Pan J-Y (2006) Fast random walk with restart and its applications. Sixth International conference on data mining. IEEE, China, pp 613–622
3. da Fontoura Costa L, Travieso G (2007) Exploring complex networks through random walks. Phys Rev E 75:16102
4. Ma T, Xia Z, Yang F (2017) An ant colony random walk algorithm for overlapping community detection. International Conference on Intelligent Data Engineering and Automated Learning. Springer, China, pp 20–26
5. Backstrom L, Leskovec J (2011) Supervised random walks: predicting and recommending links in social networks. In: Proceedings of the fourth ACM international conference on Web search and data mining. ACM, pp 635–644
6. Ghorbani M, Meybodi MR, Saghiri AM (2013) A novel self-adaptive search algorithm for unstructured peer-to-peer networks utilizing learning automata. 3rd Joint conference of AI & Robotics and 5th RoboCup Iran open international symposium. IEEE, Qazvin, Iran, pp 1–6
7. Ghorbani M, Meybodi MR, Saghiri AM (2013) A new version of k-random walks algorithm in peer-to-peer networks utilizing learning automata. 5th Conference on information and knowledge technology. IEEE Computer Society, Shiraz, Iran, pp 1–6
8. Kwok YK (2011) Peer-to-Peer computing: applications, architecture, protocols, and challenges. CRC Press, United States
9. Ghorbani M, Saghiri AM, Meybodi MR (2013) A Novel Learning based Search Algorithm for Unstructured Peer to Peer Networks. Tech J Eng Appl Sci 3:145–149
10. Gkantsidis C, Mihail M, Saberi A (2006) Random walks in peer-to-peer networks: algorithms and evaluation. Perform Eval 63:241–263
11. de Guzzi Bagnato G, Ronqui JRF, Travieso G (2018) Community detection in networks using self-avoiding random walks. Physica A: Stat Mach Appl 505:1046–1055
12. Xin Y, Xie Z-Q, Yang J (2016) The adaptive dynamic community detection algorithm based on the non-homogeneous random walking. Physica A: Stat Mach Appl 450:241–252
13. Barabási A-L, Ravasz E, Vicsek T (2001) Deterministic scale-free networks. Physica A: Stat Mach Appl 299:559–564
14. Rosvall M, Bergstrom CT (2008) Maps of random walks on complex networks reveal community structure. Proc Natl Acad Sci 105:1118–1123
15. Grady L (2006) Random walks for image segmentation. IEEE Trans Pattern Anal Mach Intell 28:1768–1783
16. Malkiel BG, McCue K (1985) A random walk down Wall Street. Norton New York
17. Van Horne JC, Parker GG (1967) The random-walk theory: an empirical test. Financ Anal J 87–92
18. Kallenberg O (2017) Random measures, theory and applications. Springer
19. Narendra KS, Thathachar MA (1989) Learning automata: an introduction. Prentice-Hall
20. Khomami MMD, Haeri MA, Meybodi MR, Saghiri AM (2017) An algorithm for weighted positive influence dominating set based on learning automata. In: 4th International conference on Knowledge-Based Engineering and Innovation (KBEI), IEEE, pp 734–740
21. Khomami MMD, Rezvanian A, Meybodi MR (2018) A new cellular learning automata-based algorithm for community detection in complex social networks. J Comput Sci 24:413–426
22. Ghamgosar M, Khomami MMD, Bagherpour N, Reza M (2017) An extended distributed learning automata based algorithm for solving the community detection problem in social networks. In: Iranian Conference on Electrical Engineering (ICEE), IEEE, pp 1520–1526
23. Saghiri AM, Meybodi MR (2016) A self-adaptive algorithm for topology matching in unstructured peer-to-peer networks. J Netw Syst Manage 24:393–426

24. Saghiri AM, Meybodi MR (2017) A distributed adaptive landmark clustering algorithm based on mOverlay and learning automata for topology mismatch problem in unstructured peer-to-peer networks. Int J Commun Syst 30:e2977
25. Saghiri AM, Meybodi MR (2018) Open asynchronous dynamic cellular learning automata and its application to allocation hub location problem. Knowl-Based Syst 139:149–169

Chapter 2
Intelligent Models of Random Walk

Abstract Intelligent models of random walk are obtained from random walk algorithms and a learning element. In this chapter, the learning element is a type of learning automata which is called as variable-structure learning automata. In this chapter, this type of learning automata is used to design three intelligent models of random walk. The first model is called Intelligent K-Random Walk based on Learning Automata (*IKRW-LA*). This model is able to predict k promising links from each arbitrary node in the networks. The second model is called Self-Organized Intelligent Random Walk based on Learning Automata (*SOIRW-LA*). This model is proposed to add the learning capability to k-random walk for path prediction. In this model, we do not need to determine the value for parameter k to find the promising links. The third model is called Intelligent Random Walk based on Learning Automata (*IRW-LA*). The advantage of this model than the others models is to predict the best path of the network considering the feedback received from the network. This model is equivalent to the first model (*IKRW-LA*) when the value of parameter k is equal to one. The convergence behavior of the *IRW-LA* is also studied. All of the mentioned models are described in this chapter.

Keywords Intelligent models of random walk · Path prediction ·
Variable-structure learning automata

2.1 Theory of Learning Automata

Learning automata are classified into two groups, namely associative and nonassociative. In associative learning automata, different context vectors are offered by the environment and the Learning automata must choose an optimal action for each of the given context vectors. In nonassociative Learning automata, no context vector is given by the environment. Instead, the learning automaton only receives reinforcement signals from the environment and chooses the optimal action according to the received signals. Fixed-structure Learning automata, variable-structure learning automata, and learning automata with changing number of actions, estimator algorithms, pursuit algorithms, and continuous action set learning automata are classified under the nonassociative learning automata group. In what follows, we provide a very brief description for fixed-structure learning automata, variable-structure learning automata, and learning automata with changing number of actions.

2.1.1 Fixed-Structure Learning Automata

The learning automaton is called fixed-structure if the followings are fixed: 1. the probability of the transition from one state to another state and 2. the action probability of any action in any state. Examples of the fixed-structure LA are L_2N_2, G_2N_2, Krylov, and Krinsky [1].

The L_2N_2 automaton has 2N states and 2 actions and tries to contribute the past behavior of the system in the future decision for choosing the sequence of actions. Based on the policy of L_2N_2, it holds an account of the number of accomplishments and failures received for each action. When the number of failures exceeds the number of successes, the automaton switches from one action and selects another one. This process is done by enlarging the state space to 2N and specifying the rules of transition from one state to another. An example of this type of automaton is given in Fig. 2.1.

Unlike automaton L_2N_2, the automaton G_2N_2 encountered with unfavorable response and then, the automaton switches from state N to $N + 1$. In contrast to automaton L_2N_2, this type of automaton performs the action α_2 more than N times before selecting action α_1 again. The state transition graphs of G_2N_2 are showed in Fig. 2.2 for both favorable and unfavorable response.

This automaton has state transitions as the same as to L_2N_2 automaton when the receiving response from the environment is favorable. But, when the response of the environment is unfavorable, a current state $i(i \neq 1, N, N + 1, 2N)$ passes to the next and previous states with equal probability. The outline of the Krylov is shown in Fig. 2.3.

Fig. 2.4 shows the state transition of Krinsky automaton in which this automaton treat similar to L_2N_2, when the response from the environment is unfavorable.

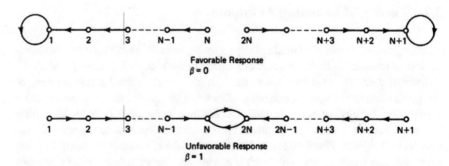

Favorable Response
$\beta = 0$

Unfavorable Response
$\beta = 1$

Fig. 2.1 The state transition graph for L_2N_2 [1]

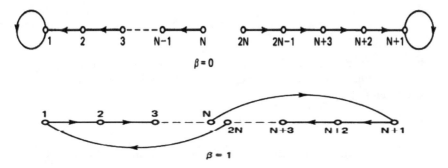

Fig. 2.2 The state transition graph G_2N_2 [1]

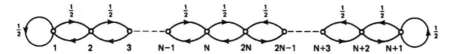

Fig. 2.3 The state transition graph of Krylov automaton [1]

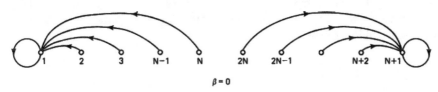

Fig. 2.4 The state transition graph of Krinsky automaton [1]

2.1.2 Variable-Structure Learning Automata

A variable-structure learning automaton is defined by the quadruple $\langle \alpha, \beta, p, T \rangle$ where α represents the action set of the learning automaton, β represents the input set, $p = \{p_1, p_2, \dots, p_r\}$ represents the action probability set, and finally, $p(k + 1) = T(\alpha(k), \beta(k), p(k))$ represents the learning algorithm. At the time instant k, this learning automaton operates as follows. Based on the action probability set $p(k)$, learning automaton randomly selects an action $\alpha_i(k)$ and performs it on the environment. After receiving the environment's reinforcement signal $\beta(k)$, learning automaton updates its action probability set based on Eq. (2.1) for favorable responses and Eq. (2.2) for unfavorable ones.

$$p_i(k + 1) = p_i(k) + a \cdot (1 - p_i(k))$$
$$p_j(k + 1) = p_j(k) - a \cdot p_j(k) \quad \forall j \, j \neq i. \tag{2.1}$$

$$p_i(k + 1) = (1 - b) \cdot p_i(k)$$

$$p_j(k + 1) = \frac{b}{r - 1} + (1 - b)p_j(k) \quad \forall j\, j \neq i. \tag{2.2}$$

In the above equations, a and b are reward and penalty parameters, respectively. If $a = b$, learning algorithm is called L_{R-P},[1] if $b \ll a$, it is called $L_{R\varepsilon P}$,[2] and if $b = 0$, it is called L_{R-I}.[3]

2.1.3 Learning Automata with Changing Number of Actions

In some applications, the number of possible or permissible choices, modeled by the actions of a learning automaton, varies over time. In this *LA*, at every time instant k, only a subset $V(k)$ of the actions of the *LA* is available for choice. The selection of the action subset $V(k)$ is made by an external system, which could determine permissible actions for a specific application. The *LA* chooses an action $\alpha_i \in V(k)$, according to the probability distribution vector $\hat{p}(k)$ defined over $V(k)$ by Eq. (2.3). The selected action is applied to a P-model environment and results in a binary response $\beta(k)$. Then, the *LA* computes $\hat{p}(k + 1)$ using Eq. (2.4) (for favorable responses) or Eq. (2.5) (for unfavorable responses).

$$K(k) = \sum_{i \in V(k)} p_i(k),$$

$$\hat{p}_i(k) = \text{prob}\,[\alpha(k) = \alpha_i | V(k) \text{ is the set of active actions}, \alpha_i \in V(k)] = \frac{p_i(k)}{K(k)} \tag{2.3}$$

$$\begin{aligned} \hat{p}_i(k + 1) &= \hat{p}_i(k) + a \cdot (1 - \hat{p}_i(k))\ \alpha(k) = \alpha_i \\ \hat{p}_i(k + 1) &= \hat{p}_j(k) + a \cdot \hat{p}_i(k) \qquad \alpha(k) = \alpha_i, \quad \forall j\, j \neq i \end{aligned} \tag{2.4}$$

$$\begin{aligned} \hat{p}_i(k + 1) &= (1 - b) \cdot \hat{p}_i(k) \qquad \alpha(k) = \alpha \\ \hat{p}_i(k + 1) &= \frac{b}{r-1} + (1 - k)\hat{p}_j(k)\ \alpha(k) = \alpha_i, \quad \forall j\, j \neq i \end{aligned} \tag{2.5}$$

Afterward, the action probability vector of the *LA* is updated using Eq. (2.6).

$$\begin{aligned} p_j(k + 1) &= \hat{p}_j(k + 1) \cdot K(k)\ \forall j,\ \alpha_j \in V(k) \\ p_j(k + 1) &= p_j(k) \qquad\qquad\qquad \forall j,\ \alpha_j \notin V(k) \end{aligned} \tag{2.6}$$

---

[1] Linear Reward Penalty.
[2] Linear Reward Epsilon Penalty.
[3] Linear Reward Inaction.

2.1.4 K-Select Action Learning Automata (KSALA)

In some applications, we need to a decision maker which selects k action in a parallel fashion. *KSALA* refers to a learning automaton in which k actions are selected instead of one action. *KSALA* was reported in [2] for solving a management problem of sensor network. The environment returns the responses to the selected actions. The responses will be used to update the probability vector of the learning automaton.

2.2 Intelligent Models of Random Walk Based on Learning Automata

In this section, three intelligent models of random walk based on learning automata are given. All models are designated based on the three phases but different algorithms are suggested for each model (Fig. 2.5). The models are described in the next three subsections.

2.2.1 Model A: Intelligent K-Random Walk Based on Learning Automata (IKRW-LA)

In k-random walk, we choose a random node in the network and trigger k-random neighbors of that node. All of triggered nodes repeat this process. In this algorithm, there is no mechanism to predict the promising neighbors with respect to the application. In this section, we propose an intelligent model of random walk utilizing *KSALA*. In this model, each node utilizes a *KSALA* to trigger the k promising nodes of its neighbors. In the rest of this section, the structure of this model is described.

An *IKRW-LA* is a dynamic graph whose structure changes with time. In this graph, each node is equipped with a *KSALA*. The *KSALA* of each node is used to select k of the best neighboring node. The *KSALA* of each node has an extra action called "self-select" which leads to select that node. For example, if a node ($node_i$) has two neighbors($node_j$ and $node_k$) then its *KSALA* has three actions ($action_j$, $action_k$, and "self-select"). Starting node is determined by the application. Upon the selection of a node, *IKRW-LA* executes a process which has three phases: paths selection, paths evaluation, and automata updating. These three phases are described below.

- **Paths selection phase**: During this phase, a set of paths is created utilizing the proposed model of random walk. The first node of a path is called as starting node. All of the paths have the same starting node. The starting node is determined by the application. The algorithm used for constructing the set of paths is described as follow. The *KSALA* resided in the starting node determines the k next nodes based on its probability vector. The *KSALA* resided in the next nodes are used to determine other nodes based on their probability vectors. The procedure of selecting the next

Fig. 2.5 The phases of the
proposed models

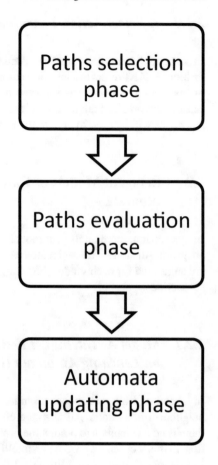

nodes based on *KSALA* is continued, until the constructed set of paths is acceptable
by the application. Note that, the conditions used for acceptance of the set of paths
considering the goal of the application are defined in this phase. Figure 2.6 shows a
snapshot of the path selection phase for 1-random walk. In this figure, dashed links
denotes the selected path. In this snapshot, the paths selection phase is finished
when a path between the source node and a destination node is found. After this
phase, the paths evaluation phase is started.

- **Paths evaluation phase**: In this phase, a reinforcement calculator function is used
  to generate the reinforcement signals. This function takes each path constructed
  in paths selection phases and user-defined metrics, and then returns a reinforcement signal for the learning automata corresponding to that path. For example,
  if the sequence of the nodes is preferable considering the user metrics, then the
  reinforcement signal is set to one (reward). When this phase is done, the automata
  updating phase is started.

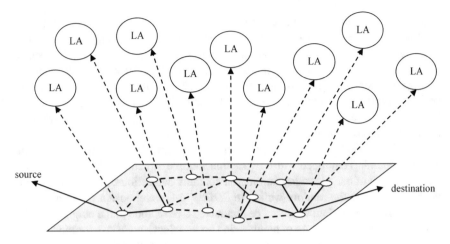

Fig. 2.6 A snapshot of path selection phase for 1-random walk

- **Automata updating phase**: During this phase, the reinforcement signal generated for each path is used to collectively update the learning automata corresponding to that path.

Remark 2.1 In the path evaluation phase, a database is used to store some information about the paths corresponding to the sequence of nodes found during the paths selection phase. For example, the favorable paths corresponding to the user-defined metrics during iterations of the algorithm are stored in the database. This database may be used to improve the accuracy of the reinforcement calculator in the path evaluation phase. Note that, the neighboring nodes of each node is determined by the application.

Remark 2.2 This model tries to predict k promising links from the starting node considering feedback received from the network.

Remark 2.3 Note that, In an iteration of the algorithm, a learning automaton LA_i may activate LA_j. Therefore, the action corresponding to LA_i will be disabled in LA_j.

2.2.2 Model B: Self-organized Intelligent Random Walk Based on Learning Automata (SOIRW-LA)

As it was previously mentioned, in k-random walk, we choose a random node in the network and trigger k-random neighbors of that node. Then, all of the triggered nodes repeat this process. The main problem of this process is to find the optimum value of the k in k-random walk in every arbitrary network. Choosing a value of k larger than the average number of neighbors of all nodes of the network results in observing useless parts of the network. For example, in a computer network, a large value for k results in generating more traffic in computer network, because it works like flooding search methods. On the other hand, if k is smaller than the average number of neighbors of all nodes of the network, the probability of selecting desirable parts of the network is decreased. For example, in a computer network, a low value for k results in generating low traffic and limited observation which leads to miss the appropriate nodes.

An *SOIRW-LA* is a dynamic graph whose structure changes with time. In this graph, each link is equipped with a two action variable-structure learning automaton. The action set of each learning automaton contains two actions: "block the random walk" and "allow the random walk". In *SOIRW-LA*, we choose a random node in the network. The random walk can be conducted from a node to the neighbor of that node. Note that, the structure of neighboring nodes of each node is determined by the application. In *SOIRW-LA*, each link between a node and the neighbors of that node is equipped with a learning automaton. The decision made by the learning automaton of each link determines whether the random walk can be conducted using that link or not. More details about the evolution of *SOIRW-LA* are given below.

Upon the selection of a node, *SOIRW-LA* executes a process which has three phases: paths selection, paths evaluation, and automata updating. These three phases are described below.

- **Paths selection phase**: During this phase, a set containing two sets of paths which are called as blocked paths and nominated paths is created. The algorithm used for creating these paths is described below.

  - The first node of a nominated path is called as starting node. All of the nominated paths have the same starting node. The starting node is determined by the application. The algorithm used for constructing the set of nominated paths is described as follows. The learning automata resided in the links of the starting node are activated to make their decisions. If the learning automaton between a node and the neighboring node of that node selects action "allow the random walk" that neighboring node will be activated. In other words, the decisions of the learning automata of the links of a node determine those neighbors which are suitable for triggering. It is obvious that, in this phase, we do not use a fixed number such as parameter k for path selection phase. The procedure of selecting the next nodes based on learning automata of the links is continued, until the constructed set of nominated paths is acceptable by the application. Note that, the conditions used for acceptance of the set of nominated paths considering the goal of the application are defined in this phase.
  - Every nominated path which was blocked by the decision of a learning automaton will be classified as blocked paths.

  After this phase, the paths evaluation phase is started.

- **Paths evaluation phase**: In this phase, a reinforcement calculator function is used to generate the reinforcement signals. This function takes each path constructed

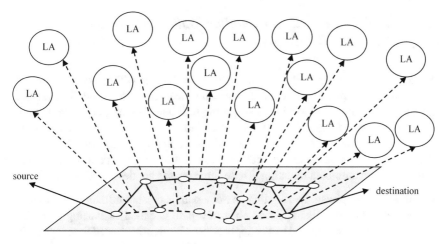

Fig. 2.7 A snapshot of the path selection phase for *SOIRW-LA*

in paths selection phases and user-defined metrics, and then returns a reinforcement signal for the learning automata corresponding to that path. For example,
if the sequence of the nodes is preferable considering the user metrics then the
reinforcement signal is set to one. In this phase, we should determine a policy for
calculating the reinforcement signals for those learning automata which blocked
the random walk. Figure 2.7 shows a snapshot of the path selection phase. In
this figure, dashed links denotes the selected paths. When this phase is done, the
automata updating phase is started.

• **Automata updating phase**: During this phase, the reinforcement signal generated
  for each path is used to collectively update the learning automata corresponding
  to that path.

Remark 2.4 In this part, an intelligent and self-organized version of random walk
is reported which have two benefits are described as follows. The first benefit of
this model is that we do not need to determine the value of k for k-random walk
algorithm and each node can conduct the random walk in self-adaptive manner. The
second benefit of this model is that *SOIRW-LA* is based on adaptive selecting of
paths according to each node's feedback. By applying learning automata for each
link in the network, those neighbors with the highest probability value of appropriate
walk considering user-based metrics in the past iterations are selected adaptively to
continue the paths selection.

Remark 2.5 In comparison with model A, this model tries to predict variable promising paths from the starting node considering feedback received from the network.

2.2.3   Model C: Intelligent Random Walk Based on Learning Automata (IRW-LA)

An *IRW-LA* is a dynamic graph whose structure changes with time. In this graph, each node is equipped with a variable-structure learning automaton. The learning automaton of each node is used to select the best neighboring node. The learning automaton of each node has an extra action called "self-select" which leads to select that node. The starting node is determined by the application. Upon the selection of a node, *IRW-LA* executes a process which has three phases: path selection, path evaluation, and automata updating. These three phases are described below.

- **Path selection phase**: During this phase, a path is created utilizing the proposed model of random walk. The first node of a path is called as starting node. The starting node is determined by the application. The algorithm used for constructing the path is described as follows. The learning automaton resided in the starting node determines the next node based on its probability vector. The learning automaton resided in the next node determines other node based on its probability vector. The procedure of selecting the next nodes based on learning automaton is continued, until the constructed path is acceptable by the application. Note that, the conditions used for acceptance of the set of path considering the goal of the application are defined in this phase. After this phase, the paths evaluation phase is started.
- **Path evaluation phase**: In this phase, a reinforcement calculator function is used to generate the reinforcement signal. This function takes the path constructed in path selection phases and user-defined metrics, and then returns a reinforcement signal for the learning automata corresponding to that path. For example, if the sequence of the nodes is preferable considering the user metrics, then the reinforcement signal is set to one. When this phase is done, the automata updating phase is started.
- **Automata updating phase**: During this phase, the reinforcement signal generated for each path is used to collectively update the learning automata corresponding to that path.

Remark 2.6 This model is equivalent to model A when the value of parameter k is equal to 1.

Remark 2.7 In comparison with model A, this model tries to predict only one promising path from the starting node considering feedback received from the network.

2.3   Performance Metrics

In order to study the performance of the proposed intelligent models of random walk, a metric called entropy is described in this section.

Entropy. Entropy is introduced into reinforcement learning as a measure of the learning process [3]. Entropy is a fundamental concept in thermodynamics, representing the degree of order or disorder in thermodynamic system, and plays an important role in various fields of computer science such as learning. Entropy essentially estimates the measure of uncertainty in the actions of learning automata during the paths selection phase. A larger value of entropy, leading to more uncertainty in the probability vector for a learning automaton, means that the automaton contains no useful information for achieving a goal and its performance is more or less similar to a random selection. The entropy at any given time is defined by Eq. (2.7)

$$h = \sum_{i=1}^{n} e_i \tag{2.7}$$

where e_i is the entropy of ith learning automaton of the *IRW-LA*. e_i is the *entropy* of learning automaton LA_i defined in Eq. (2.8) as given below

$$e_i = -\sum_{j=1}^{r_i} p_j \log(p_j) \tag{2.8}$$

where r_i is the number of actions of learning automaton LA_i. p_j is the probability of selecting action α_j of learning automaton LA_i. In a learning automaton, the entropy value is maximized when the probabilities of all actions are equal to each other. During the execution of the algorithm, the entropy value changes. Entropy can be used to study the changes occur in the sequences of walks. The value of zero for entropy means that the LAs of the nodes no longer change their actions. Higher values of entropy mean higher rates of changes in the actions selected by LAs resided in the nodes.

2.4 Convergence Behavior of *IRW-LA*

In this section, we study the convergence results of the *IRW-LA*-based algorithms, when all learning automata use L_{R-I} learning algorithm and every learning automaton has r actions. For this purpose, using the weak convergence theorems, the proposed algorithm is first approximated by an Ordinary Differential Equation (ODE). Then, it is shown that the resulting ODE converges to the solution of the optimization problem. It should be noted that we utilized the type of analyses reported in [4]. Before approximating ODE, we present some definitions and preliminary lemmas. In this part, we assume that the reward probabilities of appropriate actions of the learning automata approach to one.

Definition 2.1 The configuration of the *IRW-LA* at iteration t is defined as $Y(t) = \langle \underline{P}(t), S(t) \rangle$ where

- $\underline{P}(t) = (P_1(t), P_2(t), \ldots, P_n(t))^{\mathrm{T}}$ where $P_i(t) = (p_{i1}(t), p_{i2}(t), \ldots, p_{ir}(t))^{\mathrm{T}}$ in which $p_{ij}(t)$ is the probability of selecting action α_j of the learning automaton LA_i at iteration t. Each learning automaton has r actions.
- $\underline{S}(t) = (s_1(t), s_2(t), \ldots, s_n(t))^{\mathrm{T}}$ where $s_i(t)$ is equal to identifier of the action chosen by LA_i if the LA_i is activated to choose an action.

The initial configuration of the *IRW-LA* is denoted by $Y(0) = \langle \underline{P}(0), \underline{S}(0) \rangle$. As it was previously mentioned, upon activating the cells, a process takes the configuration, and then updates the configuration of the *IRW-LA*. The evolution of *IRW-LA* can be described by the sequence $\{Y(t)\}_{t \geq 0}$.

Before we study the convergence behavior of the *IRW-LA*, we need to define the following items.

- C denotes the set of all possible configurations of the *IRW-LA*.
- $\Delta \underline{P}(k) = E[\underline{P}(k+1)|\underline{P}(k)] - \underline{P}(k)$ in which $\Delta p_{ij}(k) = E[p_{ij}(k+1)|\underline{P}(k)] - p_{ij}(k)$. $\Delta p_{ij}(k)$ denotes the drift of the jth component of the action probability vector of automaton A_i, which is defined as the increment in the conditional expectation of p_{ij}.

Lemma 2.1 *The drift $\Delta p_{ij}(k)$ for the proposed learning algorithm can be defined as $\Delta p_{ij}(k) = a p_{ij}(k) \sum_{q \neq j} p_{iq}(k) \left(\frac{\partial f}{\partial p_{ij}} - \frac{\partial f}{\partial p_{iq}} \right)$ where $f(\cdot)$ is defined as*

Maximize $f(\underline{P}) = E[\beta|\underline{P}]$

Subject to $p_{ij} \geq 0$; $1 \leq i \leq n$ and $1 \leq j \leq r$

$$\sum_{j=1}^{r} p_{ij} = 1; \quad 1 \leq i \leq n$$

Proof Since $\{\underline{P}(k)\}_{k \geq 0}$ is a Markov process whose dynamics depend on a, and $\beta(k)$ depends only on $\underline{P}(k)$ and not on k explicitly, then $\Delta \underline{P}(k)$ can be given by a function of $\underline{P}(k)$. Now utilizing the L_{R-I} algorithm, the components of $\Delta \underline{P}(k)$ can be obtained as follows.

$$\Delta p_{ij}(k) = a p_{ij}(k) \sum_{q \neq j} p_{iq}(k) \big[E[\beta_{ij}(k)] - E[\beta_{iq}(k)] \big], \qquad (2.9)$$

From the definition of $f(\cdot)$, we have

$$f(\underline{P}) = \sum_{q \neq j} p_{iq}(k) E[\beta_{iq}(k)]. \qquad (2.10)$$

Differentiating both sides of Eq. (2.10), we obtain

$$\frac{\partial f}{\partial p_{ij}} = E[\beta_{iq}(k)]. \qquad (2.11)$$

Substituting Eq. (2.9) in Eq. (2.11), we have

$$\Delta p_{ij}(k) = a p_{ij}(k) \sum_{q \neq j} p_{iq}(k) \left(\frac{\partial f}{\partial p_{ij}} - \frac{\partial f}{\partial p_{iq}} \right) \tag{2.12}$$

and hence the proof of the lemma.

Proposition 2.1 *It can be seen that Δp_{ij} is not directly a function of the time step k, and so it can be rewritten as*

$$\Delta p_{ij}(k) = a y_{ij}\big(\underline{P}(k)\big) \tag{2.13}$$

where $y_{ij}\big(\underline{P}\big) = p_{ij} \sum_{q \neq j} p_{iq} \left(\frac{\partial f}{\partial p_{ij}} - \frac{\partial f}{\partial p_{iq}} \right)$. For each $a > 0$, $\{\underline{P}(k)\}_{k \geq 0}$ is a Markov process whose dynamics depend on a, and can be described by the following difference equation:

$$\underline{P}(k+1) = \underline{P}(k) + a\mathbb{G}\big(\underline{P}(k), \underline{S}(k), \underline{\beta}(k)\big) \tag{2.14}$$

where $\underline{S}(k)$ denotes the set of actions selected by the learning automata at instant k, $\underline{\beta}(k) = (\beta_1(k), \beta_2(k), \ldots, \beta_n(k))$ denotes the set of reinforcement signals emitted from the environment in response to the chosen actions, and $\mathbb{G}(., ., .)$ is the learning function by which the action probabilities are updated.

Define a piecewise-constant interpolation $\underline{P}^a(t)$ of $\underline{P}(k)$ as

$$\underline{P}^a(t) = \underline{P}(k) \quad \text{if} \quad t \in [ka, (k+1)a] \tag{2.15}$$

where a is the learning parameter used in Eqs. (2.1) and (2.2). Now consider the sequence $\{\underline{P}^a(\cdot) : a > 0\}$. We are interested in the limit of this sequence as a converges to zero. The following theorem gives the limiting behavior of $\underline{P}^a$ as $a \to 0$.

Theorem 2.1 *Given the sequence of interpolated processes $\{\underline{P}^a(\cdot) : a > 0\}$ and $X_0 = \underline{P}^a(0) = \underline{P}(0)$. The sequence $\{\underline{P}^a(\cdot)\}$ weakly converges to $\underline{X}(\cdot)$ as a converges to zero, where $\underline{X}(\cdot)$ is the solution of the following ODE:*

$$\frac{dX_{ij}}{dt} = l_{ij}\big(\underline{X}\big), \underline{X}(0) = \underline{P}(0) \tag{2.16}$$

Proof The proof of this theorem is based on weak convergence theorem. The following conditions are satisfied by the learning algorithm given in Eqs. (2.1) and (2.2).

- $\{\underline{P}(k), \big(\underline{S}(k-1), \underline{\beta}(k-1)\big)\}_{k \geq 0}$ is a Markov process.
- $\big(\underline{S}(k-1), \underline{\beta}(k-1)\big)$ take values in a compact metric space. Note that, the reinforcement signals take values from the closed interval $[0, 1]$.

- The function $\mathbb{G}(.,.,.)$ defined in Eq. (2.14) is bounded. In addition, this function is continuous and independent of a.
- If $\underline{P}(k) = \underline{P}$ is a constant, then $\left\{\left(\underline{S}(k), \underline{\beta}(k)\right)\right\}_{k \geq 0}$ is an independent identically distributed sequence.
- The ODE given in Eq. (2.16) has a unique solution for each initial condition $\underline{X}(0)$.

Hence, using the weak convergence theorem, the sequence $\{\underline{P}^a(.)\}$ converges weakly as $a \to 0$ to the solution of the following ODE:

$$\frac{dX_{ij}}{dt} = l_{ij}(\underline{X}), \underline{X}(0) = \underline{P}(0) \tag{2.17}$$

where $l_{ij}(\underline{X})$ is determined by $E^P \mathbb{G}(P(k), S(k), \beta(k))$, and E^P denotes the expectation with respect to an invariant measure. Since for $\underline{P}(k) = \underline{P}$, the sequence $(s(k), \beta(k))$ is an independent identically distributed sequence whose distribution depends only on $\underline{P}$ and the rule under which the selected actions are rewarded. Therefore, for the value of every component such as l_{ij} is defined as given in Eq. (2.12), and hence the theorem is complete.

Theorem 2.2 *For large values of k and small enough values of learning rate a, the asymptotic behavior of $\underline{P}(k)$ generated by the IRW-LA can be well approximated by the solution to the ODE given in Eq. (2.17) with the same initialization.*

Proof The interpolated process $\{\underline{P}^a(t)\}_{t \geq 0}$ is a sequence of random variables. These random variables take values from $D^{m_1 \times m_2 \times \cdots \times m_n}$, where $D^{m_1 \times m_2 \times \cdots \times m_n}$ is the space of all functions that, at each point, are continuous on the right and have a limit on the leftover $[0, \infty)$ and take values from a bounded subset of $\Re^{m_1 \times m_2 \times \cdots \times m_n}$. Let $v_T(\cdot)$ be a function over $D^{m_1 \times m_2 \times \cdots \times m_n}$ and is given by

$$v_T(\underline{Y}) = \sup_{0 \leq t \leq T} \|\underline{W}(t) - \underline{X}(t)\| \tag{2.18}$$

for every $T < \infty$, we must show that with probability increasingly close to one as a decreases, $\underline{P}(k)$ follows the solution of the ODE given in Eq. (2.17), with an error bounded above by some fixed $\varepsilon > 0$. This result can be specialized to characterize the long-term behavior of $\underline{P}(k)$, when the initial configuration, $\underline{P}(0)$, is in the neighborhood of an asymptotically stable compatible configuration. Let $\underline{P}^0$ be the equilibrium point to which the solution of the ODE given in Eq. (2.17) when the initial condition is $\underline{P}(0)$.

Using the weak convergence result, we have

$$E[v_T(\underline{P}^a)]_{a \to \infty} \to E[v_T(\underline{X})] \tag{2.19}$$

Note that, $\underline{X}$ is the solution to the ODE defined in Eq. (2.17). Let $\underline{P}^0$ be the equilibrium point to which the solution of the ODE converges, when $\underline{X}(0) = \underline{X}_0$ is the used initial condition. The weak convergence result given in Eq. (2.19) along with the nature of interpolation given in Eq. (2.15) imply that for the given initial configuration, any $\varepsilon > 0$ and integers k_1 and k_2, where $0 < k_1 < k_2 < \infty$, there exists a a^* such that

$$\text{Prob}\left[\sup_{k_1 \leq k \leq k_2} \left\| \underline{P}(k) - \underline{P}^0 \right\| > \varepsilon \right] = 0 \quad \forall a < a^* \tag{2.20}$$

where $\underline{P}^0$ is an asymptotically stable equilibrium point of the ODE given in Eq. (2.17). Note that, for all initial configurations in small neighborhood of $\underline{P}^0$, the learning automata converges to $\underline{P}^0$, and the proof is complete.

Proposition 2.2 *In an IRW-LA, if conditions mentioned in Theorems 2.1 and 2.2 is satisfied and $\varepsilon \to 0$ by proper choice of parameters in the learning automata, $\lim_{t \to \infty} (h(t)) = h^0$ in which h^0 is a constant value.*

Proof By expanding $\lim_{t \to \infty} (h(t))$, we have $\lim_{t \to \infty} \left(-\sum_{k=1}^{n} \sum_{l=1}^{r_k} \left[p_{kl}(t) \times \ln(p_{kl}(t)) \right] \right)$. According to the result of Theorems 2.1 and 2.2, we have $\lim_{t \to \infty} (\underline{P}(t)) = \underline{P}^0$. Therefore, $h(t)$ approaches to $-\sum_{k=1}^{n} \sum_{l=1}^{r_k} \left[p_{kl}^0(t) \times \ln(p_{kl}^0(t)) \right]$ and the proof is completed.

References

1. Narendra KS, Thathachar MA (1989) Learning automata: an introduction. Prentice-Hall
2. Abolhassani M, Esnaashari M, Meybodi MR (2009) LADIT: A learning automata fault tolerant routing protocol for sensor networks. Yazd, Iran, pp 12–18
3. Rezvanian A, Saghiri AM, Vahidipour M, Esnaashari M, Meybodi MR (2018) Recent advances in learning automata. Springer
4. Akbari Torkestani J, Meybodi MR (2010) Mobility-based multicast routing algorithm for wireless mobile Ad-hoc networks: a learning automata approach. Comput Commun 33:721–735

Chapter 3
Applications

Abstract All of the intelligent models of random walk proposed in Chap. 2 are domain independent. Therefore, these models can be applied in a wide variety of problems. As it was previously discussed, all of the proposed models can be used to design prediction models based on random walk algorithms. This characteristic can be used to design problem-solving methods in large-scale systems such as peer-to-peer networks and social networks. Therefore, the proposed models will be used to solve two problems in peer-to-peer networks and also one problem in social networks. In peer-to-peer networks, we focus on designing intelligent search algorithms which lead to find a specific object in the network. In social networks, we focus on designing an intelligent mechanism for finding a set of nodes called PIDS which leads to solve influence maximization problem. In the rest of this section, we give the required information about the selected problems, and then several solutions based on intelligent models of random walk are studied.

Keywords Intelligent models of random walk · Prediction models · Peer-to-Peer networks · Intelligent search algorithms · Social networks · Influence maximization

3.1 Basic Concepts

In this section, the required information from peer-to-peer networks and social networks are given. In the first subsection, we focus on the problem of designing search algorithms in peer-to-peer networks and in the second subsection, we study the problem of designing influence maximization algorithms in social networks.

3.1.1 Peer-to-Peer Networks

A peer-to-peer network is a network of computers in which all peers have the same role (client and server). In peer-to-peer networks, all computers communicate directly with each other. In a peer-to-peer overlay network, the network is constructed over

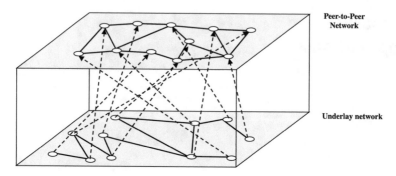

Fig. 3.1 A snapshot of a peer-to-peer network [3]

another network called underlay network (Fig. 3.1). When these networks are established over the internet, the resources of each peer can be shared in the whole network. In recent years, many instances of these networks such as Skype [1], BitTorrent [1], and Bitcoin [2] are widely used by the users.

Peer-to-peer networks can be classified as structured and unstructured. *Gnutella* [4] and *Freenet* [5] are two well-known instances of unstructured peer-to-peer networks. In these networks, there are some lightweight algorithms to manage the topology of the peer-to-peer network. These networks are widely used because their design is simple. In the other hand, we have the structured peer-to-peer networks. In these networks such as *Chord* [6] and *CAN* [7], the distributed algorithms are designated to manage the topology of the peer-to-peer network that can support efficient resource discovery algorithms. The discovery of the objects in unstructured peer-to-peer networks is a challenging problem because there is no central mechanism to manage the resources.

Nowadays, most of the applications of peer-to-peer networks focus on unstructured peer-to-peer networks, and therefore designing an efficient search algorithm is the most important management problem. Random walk-based algorithms are used to design a type of search algorithms [8–10]. In the literature, reinforcement learning techniques are utilized to improve search efficiency in peer-to-peer networks [11]. In these algorithms, each peer tries to learn required information about network status, and then makes the decision for the next episode of search based on the previous feedbacks.

3.1.2 Social Networks

Social networks have become very popular in recent years because of the increasing proliferation and affordability of Internet-enabled devices such as personal computers, mobile devices, and other more recent hardware innovations such as Internet tablets. In general, a social network is defined as a network of interactions or rela-

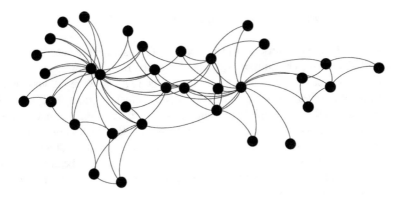

Fig. 3.2 An example of social network graphs

tionships, where the nodes consist of actors, and the edges consist of the relationships or interactions between these actors. A generalization of the idea of social networks is that of information networks, in which the nodes could comprise either actors or entities, and the edges, represent the relationships between them. Clearly, the concept of social networks is not restricted to the specific case of an Internet-based social network such as Facebook; the problem of social networking has been applied in many domains such as sociology in terms of generic interactions between any groups of actors. Such interactions may be in any conventional or nonconventional form, whether they be face-to-face interactions, telecommunication interactions, email interactions, or postal mail interactions. A number of important problems arise in the context of structural analysis of social networks. An important line of research is to try to understand and model the nature of very large online networks (Fig. 3.2). For more information about social networks graph please see [12].

3.1.2.1 Random Walks and Their Applications in Social Networks

Ranking is one of the most famous methods in web search. Starting with the PageRank algorithm [13] for ranking web documents, the broad principle can also be applied for searching and ranking entities and actors in social networks. The PageRank algorithm uses random walk techniques for the ranking process. The idea of random walk approach is applied to the network in order to estimate the probability of visiting each node. This probability, which is achieved by the random walk, is estimated as the PageRank. Clearly, the nodes which are structurally well connected have a higher PageRank, and naturally of greater importance. Moreover, random walk techniques can be viewed as a tool to personalize the PageRank computation process, by biasing the ranking toward particular kinds of nodes.

3.1.2.2 Community Detection in Social Networks

One of the most important problems in the context of social network analysis is that of community detection. The community detection problem is closely associated with clustering in networks, and its attempts to determine regions of the network, which are dense in terms of the linkage behavior. The community detection problem implies to the generic problem of graph partitioning [14] which partitions the network into dense regions based on the linkage behavior. However, social networks are usually dynamic, and this leads to some unique issues from a community detection point of view. In some cases, it is also possible to integrate the content behavior into the community detection process. In this case, the content is applied to determine groups of actor with similar interests (Fig. 3.3).

Influence maximization in social networks. In many applications, some nodes in a social network have labeled, and it may be desirable to use to carry the structural information in the social network in order to propagate these labels. Social networks also include valuable knowledge about the content and structure of the network, which may be applied for this goal. For example, within the context of the drinking problem, a person who is called as drinker can be converted to an abstainer through intervention program and have a positive impact on his direct friends (called neighbors). However, he/she might turn back into a binge drinker and have a negative impact on his neighbors if many of his friends are binge drinkers. Our idea is to train every binge drinker since this will diminish the likelihood of converted binge drinker being transformed through his binge drinker friends who are not contributing in the intervention program. On the other hand, due to the budget limitations, it is impossible to include all the binge drinkers in the intervention program. This problem can be interpreted and motivated as the Positive Influence Dominating Set (PIDS) in the social network. Let $G\langle V, E\rangle$ be the graph of a network in which V is the set of nodes and E is the set of edges. For such a network, a PIDS is a subset $P \subset V$ with

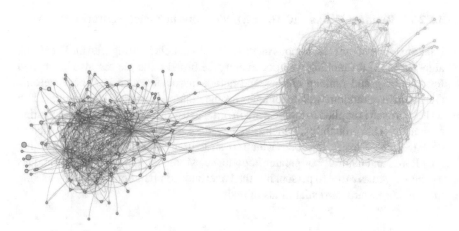

Fig. 3.3 A typical example of community detection [15]

Fig. 3.4 Example of
minimum positive influence
dominating set in a typical
input network

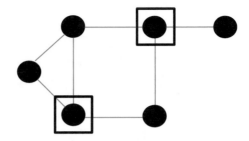

a minimum size such that any node $v \in V$ is dominated by at least $\left(\frac{d(v)}{2}\right)$ nodes in P where d(v) is the degree of node v. The Minimum Positive Influence Dominating Set (MPIDS) is a PIDS with minimum cardinality. MPIDS has many applications in online social networks. For example, in terms of viral marketing [16], an influential individual can efficiently change the decisions by other people to adopt or refuse a product or products. Due to structural changes in positive effects on social networks, it can be guaranteed through the constructing of an MPIDS that each node has more positive neighbors than negative ones. Figure 3.2 gives a typical example. The black nodes represent one of the MPIDS in this network (Fig. 3.4).

3.2 Application of Model A in Peer-to-Peer Networks

In k-random walk-based search algorithm for unstructured peer-to-peer networks such as those reported in [8, 9], it utilizes an adaptive mechanism for selecting k appropriate neighbor which affects the performance of the search algorithm. Therefore, in this section, model A is used to design a search algorithm in peer-to-peer networks called as I-Walker algorithm. This algorithm is a general version of a learning automata search algorithm which was reported in [9]. In I-Walker algorithm, an intelligent search algorithm has been developed to overcome the challenge of finding k appropriate neighbors. In the rest of this section, the structure of the proposed algorithm is described and then, the results of the simulations are reported.

3.2.1   I-Walker Algorithm

In this section, we propose a search algorithm based on *IKRW-LA* for unstructured peer-to-peer network. In this algorithm, we use an *IKRW-LA* whose structure is isomorphic to the peer-to-peer network. Figure 3.5 shows a snapshot of the execution of the proposed algorithm when the starting node decides to use its corresponding learning automaton. Note that, each node is mapped to a learning automaton. Each peer of the peer-to-peer network corresponds to a node of *IKRW-LA*. In the proposed

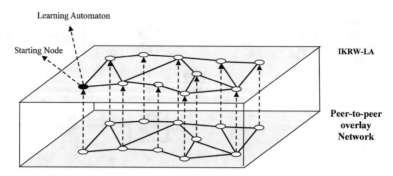

Fig. 3.5 Mapping *IKRW-LA* to peer-to-peer network

algorithm, evolving the *IKRW-LA* results in conducting an intelligent version of random walk on the peer-to-peer network which leads to executing the search algorithm. The detailed description of the proposed algorithm is given in the rest of this section.

Once the *IKRW-LA* is created, the proposed algorithm utilizes it to manage the search process. Upon issuing a query by a peer (peer$_i$) of the network, a process is executed by the peer which consists of three phases: *Paths selection* Phase, *Paths evaluation* phase, and *Automata Updating* phase. These phases are briefly described as below.

- **Paths selection phase**. During this phase, several paths are created utilizing the proposed model of random walk algorithm over the peer-to-peer network. These paths are equivalent to the paths for forwarding the query in the peer-to-peer network. The first node of the paths is called as starting node and the starting node is equivalent to a peer which issued a query. The algorithm executed via a path is explained as follows. Every time a peer receives query, it searches about the related object in its local database. In the peer, if the object is not found, locating the object will continue via query forwarding to a set of neighboring peers selected by decision made by the learning automaton associated with that peer. If the object is found, locating the object will be terminated and a message containing the result of the search will be sent to all peers which participated in the query forwarding. In the worst case, if the object is not found and the length of the path is higher than the value of parameter max_length (a parameter for the algorithm), then locating the object will be terminated, and a message containing the result of the search will be sent to all peers which participated in the query forwarding. When this phase is done, the paths evaluation phase is started.
- **Paths evaluation phase**. As it was previously mentioned in part 2, in this phase, a reinforcement calculator function is used to generate the reinforcement signal. This function takes the path constructed in path selection phases and return 1 (as reward) if the appropriate object is found during the search process and 0 (as

punishment) otherwise. When this phase is done, the automata updating phase is started.

- **Automata updating phase**. During this phase, the reinforcement signal generated for each path is used to collectively update the learning automata corresponding to that path.

3.2.2 Simulation Results

In this section, we give the simulation results. In the rest of this section, we give some information about the simulator, and then we report the results of the simulation. We use *OverSim* simulator [17]. This simulator is a framework in *OMNet++* for simulating peer-to-peer systems. A random graph with 1000 nodes is used to constructed the topology of the network. The average out-degree of each peer is 3. 100 distinct objects are randomly distributed among various peers. The *TTL* of the queries is set to 6.

Experiment 1

This experiment is conducted to show the impact of the parameter k on the performance of the proposed algorithm with respect to the success rate. For this purpose, the proposed algorithm is tested for 15 values for parameter k = 1–15. Figure. 3.6 shows our results of simulations with regard that k is the number of the walkers in different algorithms. In Fig. 3.6, we found that the proposed algorithm has a high success rate even for small k values. In random walk due to random selection of walkers, it has a low success rate.

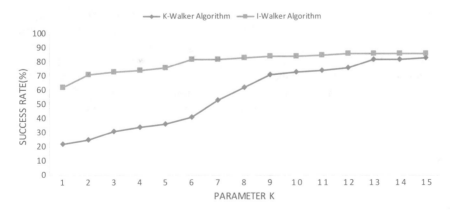

Fig. 3.6 Success rate versus parameter k

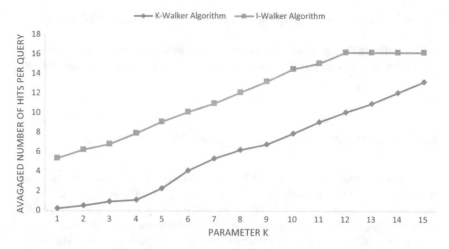

Fig. 3.7 Averaged number of hits per query versus number of walkers

Experiment 2

This experiment is conducted to show the impact of the parameter k on the performance of the proposed algorithm with respect to the averaged number of hits per query (Fig. 3.7). In the proposed algorithm, the selected neighbors always have the highest probability value of participation in the search. Therefore, the chance of discovering objects in the proposed algorithm when the query is propagated is higher that k-random walk algorithm.

3.2.3   Conclusion

In this section, an application of *IKRW-LA* model in peer-to-peer networks is reported. In each peer, our solution selects promising neighbors considering feedback received from the network instead of selecting random neighbors. From the technical point of view, utilizing learning automaton for each peer in the network, all the suitable peers are gradually selected to go on the search. We compared our algorithm's performance with other solutions via simulations. The simulation results show the superiority of the proposed algorithm in comparison with k-random walk algorithm.

3.3 Application of Model B in Peer-to-Peer Networks

As it was previously mentioned, in k-random walk-based search algorithm for unstructured peer-to-peer networks such as those reported in [8, 10], an adaptive mechanism is utilized for selecting k appropriate neighbor of the initiator peer which affects the performance of the search algorithm. In this section, model B is used to design a search algorithm in peer-to-peer networks called as S-Walker algorithm. This algorithm is a general version of a learning automata search algorithm which was reported in [10]. S-Walker algorithm does not utilize a fixed value of k for k-random walk algorithm and each peer issue walkers in a self-organized manner. In the rest of this section, the structure of the proposed algorithm is described and then, the results of the simulations are reported.

3.3.1 S-Walker Algorithm

In this section, we propose a search algorithm based on *SOIRW-LA* for unstructured peer-to-peer network. In this algorithm, we use a *SOIRW-LA* whose structure is isomorphic to the peer-to-peer network. Figure 3.8 shows a snapshot of the execution of the proposed algorithm when the starting node decides to use the learning automata associated to its links. Note that, each link is mapped to a learning automaton. Each peer of the peer-to-peer network corresponds to a node of *SOIRW-LA*. In the proposed algorithm, evolving the *SOIRW-LA* results in conducting an intelligent version of random walk on the peer-to-peer network which leads to executing the search algorithm. The detailed description of the proposed algorithm is given in the rest of this section.

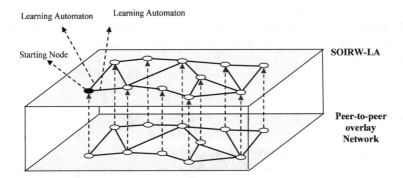

Fig. 3.8 Mapping *SOIRW-LA* to peer-to-peer network

Once the *SOIRW-LA* is created, the proposed algorithm utilizes it to manage the search process. Upon issuing a query by a peer (peer$_i$) of the network, a process is executed by the peer which consists of three phases: ***Paths selection* Phase**, ***Paths evaluation* phase**, and ***Automata Updating* phase**. These phases are briefly described as below.

- **Paths selection phase**. As it was previously mentioned, during this phase, a set containing two sets of paths which are called as blocked paths and nominated paths is created. More details about the algorithm used for creating these paths are described in part 2. These paths are equivalent to the paths for forwarding the query in the peer-to-peer network. The starting node is equivalent to a peer which issued a query. The algorithm executed via a path is explained as follow. Every time a peer receives a query, it searches about the related object in its local database. In the peer, if the object is not found, locating the object will continue via query forwarding to a set of neighboring peers selected by decision made by the learning automata associated with the links of that peer. If the object is found, locating the object will be terminated and a message containing the result of the search will be sent to all peers which participated in the query forwarding. Note that, if the path is blocked by action "block the random walk" of a learning automaton, locating the object will be terminated and a message containing the result of the search will be sent to all peers which participated in the query forwarding. In the worst case, if the object is not found, then locating the object will be terminated and a message containing the result of the search will be sent to all peers which participated in the query forwarding. When this phase is done, the paths evaluation phase is started.
- **Paths evaluation phase**. As it was previously mentioned in part 2, in this phase, a reinforcement calculator function is used to generate the reinforcement signal. This function takes a nominated path constructed in paths selection phases, and then return 1 (as reward) if the appropriate object is found during the search process and 0 (as punishment) otherwise. For blocked paths, we use different policies which are described as follows. The reinforcement calculator function takes a blocked path constructed in paths selection phases, and then return 1 (as reward) if the length of the path is higher than the value of parameter max_length (a parameter for the algorithm) and 0 (as punishment) otherwise. When this phase is done, the automata updating phase is started.
- **Automata updating phase**. During this phase, the reinforcement signal generated for each path is used to collectively update the learning automata corresponding to that path.

3.3.2 Simulation Results

The simulation setup and environment are similar to settings mentioned in Sect. 3.2.2. As mentioned before, OverSim simulator was used as a simulator. The number of simulation rounds is set to 20. In each round, all the peer executes the algorithm

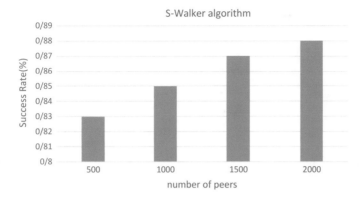

Fig. 3.9 Impact of the number of peer on the success rate of S-Walker algorithm

at the same time. In order to study the performance of the proposed algorithm, the following experiments were performed.

Experiment 1

This experiment is conducted to study the scalability of the proposed algorithm. Figure 3.9 shows that the proposed algorithm keeps its functionality as the number of peers increases.

Experiment 2

This experiment is conducted to study the overhead of the algorithm in comparison with random walk strategy. For each round of the simulation, the average number of the propagated query in each round will be calculable (Fig. 3.10). These results are shown that after fourth round which demonstrates that the average number of the propagated query decreases until it converges a stable value.

3.3.3   Conclusion

In this section, an application of *SOIRW-LA* model in peer-to-peer networks is given. Utilizing learning automaton for each link in the network, all the suitable links are gradually selected to conduct the search process. A main benefit of the proposed algorithm is that we do not need to set a value for parameter k. In each peer, our solution selects promising neighbors considering feedback received from the network instead of selecting random neighbors in self-organized manner. The simulation results show the superiority of the proposed algorithm in comparison with random walk algorithm.

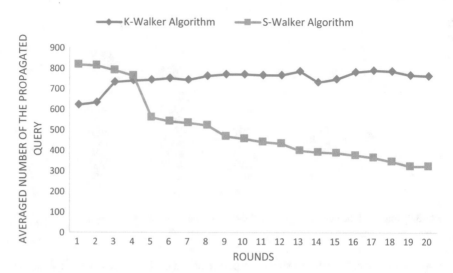

Fig. 3.10 Averaged number of propagated query

3.4   Application of Model C in Social Networks

This section provides the details of Intelligent Random Walk algorithms based on Learning Automata (*IRW-LA*) in the application of social networks. *IRW-LA* is applied for devising a search algorithm in social networks. We also studied Minimum Positive Influence Dominating Set *(MPIDS)* problem by the use of the proposed model.

3.4.1   RW-PIDS *Algorithm*

In the proposed algorithm, each node of the underlying network is initially equipped with a learning automaton. Therefore, a network of learning automata isomorphic to the structure of the social network is initially constructed. The learning automaton of each node is used to select a neighboring node. Starting node is highly depending on by the application. Upon the selection of a node, *IRW-LA* executes a process which has three phases: path selection, path evaluation, and automata updating. In the remainder of this section, these three phases are described and simulation results are reported. During the execution of the algorithm, a node is selected randomly and then, the phases of *IRW-LA* are executed based on this node called *nodes*. As it was previously mentioned, *IRW-LA* has three phases, and we adapt it for finding *MPIDS* as described below.

- **Path selection phase**. A path from the starting node *nodes* is constructed based on the learning automata based random walk. In this phase, learning automaton corresponding to a node of the path determines the next node of the path. After this phase, we go to *path evaluation* phase.

- **Path evaluation phase**. In this phase, if the nodes mentioned in the path constructed in the previous phase results in a feasible solution, the reinforcement signal (β) is set to 1 (reward) and 0 (punishment) otherwise. At the end of this phase, we go to automata updating phase.
- **Automata updating phase**. During this phase, the reinforcement signal generated by the previous phase is used to collectively update those learning automata which participated in path construction.

The process of these three phases is repeated, until a predefined threshold either iteration number (called ITR) is satisfied or the entropy of *IRW-LA* reach to a threshold (called ENT-LA). After these three processes are done, nodes mentioned in the path of the *IRW-LA* are extracted to the user as a solution.

3.4.2   Simulation Results

To show the effectiveness of the proposed model, several experiments have been devised on well-known real and synthetic networks. Based on the proposed model, if the stopping condition reaches to a satisfactory number of iterations or the value of entropy is lower than 0.1, the model stops. In experiment 1 and experiment 2, the behavior of learning model is studied. Then, we first apply the *RW-PIDS* to find *MPIDS* for a given network which leads to solve influence maximization problem. Various algorithms and heuristics have been examined for influence maximization and the obtained results are compared.

Experiment 1

One of the hard tasks in any learning algorithm is choosing the appropriate learning rate *a*. Therefore, this experiment is devised in order to study the effect of choosing various learning rates on the efficiency of the proposed algorithm. In Tables 3.1, 3.2, 3.3, 3.4, 3.5, and 3.6, the sizes of the obtained MPIDS are reported based on different learning rates and networks. Each experiment is carried out 30 times, and the results are stated based on the mean and standard deviation values. From Tables 3.1, 3.2, 3.3, 3.4, 3.5, and 3.6, we may conclude that the performance of the algorithm highly depends on the value of the learning rate *a*, and in most cases, the lower the learning rate *a* leads to the small size of the obtained result. The best achievement for each network is shown in boldface manner.

Experiment 2

In this experiment, we studied the convergence behavior of the proposed algorithm. To achieve this goal, the entropy value versus the iteration number is plotted in Fig. 3.11. By the use of the entropy, it is possible to show the uncertainty for learning automata in choosing the next group of actions; the greater value of entropy represents

Table 3.1 The minimum PIDS obtained by Algorithm 1 for different learning rates from 0.001 to 0.006 in terms of the average and standard deviation values

Learning rate	Karate	Dolphins	Jazz	Football
0.001	19.00 ± 1.84	47.08 ± 2.05	176.00 ± 2.22	89.00 ±1.55
0.002	19.00 ± 1.19	45.00 ± 2.07	183.00 ±3.31	88.00 ± 1.99
0.003	19.00 ± 1.19	45.00 ±2.11	180.00 ±4.02	90.00 ±2.01
0.004	19.00 ± 2.11	46.00 ±3.02	180.00 ±4.11	90.50 ±2.09
0.005	19.00 ± 2.12	46.00 ±3.09	181.00 ±4.91	91.00 ±1.99
0.006	19.00 ± 2.22	46.00 ±3.11	182.00 ±4.99	92.00 ±1.88

Table 3.2 The minimum PIDS obtained by Algorithm 1 for different learning rates from 0.007 to 0.03 in terms of the average and standard deviation values

Learning rate	Karate	Dolphins	Jazz	Football
0.007	19.00 ±2.23	45.00 ±3.89	180.00 ±4.55	90.00 ±2.22
0.008	19.00 ±2.25	46.00 ±3.12	181.00 ±5.01	91.00 ±2.01
0.009	19.00 ±2.33	45.00 ±4.12	180.00 ±5.05	90.00 ±2.34
0.01	19.00 ±2.35	45.00 ±4.99	179.00 ±4.25	88.28 ±5.22
0.02	19.01 ±3.51	47.00 ±3.02	178.00 ±4.33	93.00 ±3.58
0.03	19. 92 ±3.55	46.00 ±3.05	180.00 ±5.08	93.07 ±3.84

Table 3.3 The minimum PIDS obtained by Algorithm 1 for different learning rates from 0.04 to 0.1 in terms of the average and standard deviation values

Learning rate	Karate	Dolphins	Jazz	Football
0.04	19. 73 ±3.56	46.01 ±3.41	179.00 ±4.62	91.01 ±4.02
0.05	19. 29 ±3.11	46.03 ±3.59	179.00 ±4.77	99.03 ±5.77
0.06	19. 17 ±3.05	45.27 ±5.11	178.00 ±3.77	92.18 ±3.54
0.07	19.23 ±3.19	47.23 ±3.33	179.00 ±6.19	91.38 ±5.55
0.08	19.23 ±3.22	46.25 ±3.65	179.00 ±6.25	97.28 ±6.45
0.09	19.46 ±4.56	46.50 ±3.99	178.00 ±4.025	92.72 ±4.22
0.1	19.93 ±4.66	47.34 ±4.12	181.00 ±5.95	93.44 ±4.25

higher degree of uncertainty of the next action set, which means that the learning automata have no useful data for decision. On the other hand, a lower uncertainty value shows that the learning automaton selects a proper action set with a high probability.

As shown in Fig. 3.11, the entropy value continuously decreases by increasing in iteration number. We note that the soft reduction in entropy value indicates that the algorithm converges to a proper action. In addition, the trend in the reduction of the entropy value is considerably different in different networks, meaning that the entropy value depends on the topology of the networks.

Table 3.4 The minimum PIDS obtained by Algorithm 1 for different learning rates from 0.001 to 0.006 in terms of the average and standard deviation values

Learning rate	BA-100	BA-200	BA-300	BA-400	BA-500
0.001	72.00 ± 1.22	166.00 ± 2.51	242.03 ±1.21	317.00 ± 0.05	444.00 ± 5.22
0.002	72.50 ±2.14	166.00 ±3.99	242.00 ± 1.02	317.00 ±0.12	444.38 ±5.45
0.003	72.00 ±2.19	166.02 ±4.25	244.02 ±1.25	318.00 ±1.02	446.00 ±6.25
0.004	72.05 ±2.33	167.04 ±4.41	242.00 ±2.51	317.00 ±0.34	445.09 ±5.25
0.005	73.00 ±2.02	170.00 ±3.21	244.02 ±3.14	317.00 ±0.44	451.00 ±5.21
0.006	72.00 ±2.51	168.00 ±3.02	244.00 ±3.24	319.00 ±0.65	446.00 ±3.56

Table 3.5 The minimum PIDS obtained by Algorithm 1 for different learning rates from 0.007 to 0.03 in terms of the average and standard deviation values

Learning rate	BA-100	BA-200	BA-300	BA-400	BA-500
0.007	72.00 ±2.55	169.00 ±4.11	245.00 ±2.99	319.00 ±0.75	446.00 ±4.25
0.008	77.00 ±4.21	167.00 ±4.12	245.00 ±2.74	318.00 ±1.85	448.00 ±5.25
0.009	74.00 ±3.25	169.00 ±4.22	249.00 ±4.25	319.00 ±2.22	451.00 ±5.31
0.01	73.00 ±2.18	166.00 ±5.12	246.00 ±4.33	322.00 ±2.51	451.00 ±5.51
0.02	74.00 ±3.22	168.00 ±4.21	249.05 ±4.45	319.03 ±2.65	460.00 ±7.21
0.03	77.00 ±5.14	171.00 ±3.21	251.00 ±4.32	326.08 ±4.16	455.00 ±6.54

Table 3.6 The minimum PIDS obtained by Algorithm 1 for different learning rates from 0.04 to 0.1 in terms of the average and standard deviation values

Learning rate	BA-100	BA-200	BA-300	BA-400	BA-500
0.04	77.01 ±5.21	168.04 ±4.51	252.02 ±5.02	329.59 ±4.52	461.02 ±6.66
0.05	77.08 ±7.35	170.14 ±4.22	249.64 ±5.21	332.13 ±5.04	459.70 ±7.21
0.06	77.18 ±7.36	176.34 ±5.41	251.38 ±6.28	333.52 ±5.24	462.83 ±6.58
0.07	77.36 ±7.38	177.41 ±5.51	252.75 ±6.55	338.92 ±5.591	461.61 ±5.95
0.08	77.73 ±7.55	178.34 ±5.81	252.75 ±6.58	334.05 ±5.95	460.78 ±7.25
0.09	77.72 ±5.65	179.66 ±5.91	252.53 ±6.67	334.12 ±5.99	463.19 ±7.11
0.1	78.56 ±6.45	171.62 ±5.67	255.00 ±7.65	336.25 ±6.38	467.19 ±7.24

Experiment 3

To study the effectiveness, we designed numerous experiments to compare the influence spread achieved by our proposed algorithm with High-degree [18], PageRank [19], HITS [20], CELF [21], and CELF++ [22]. The computational configurations below were set by applying to these algorithms:

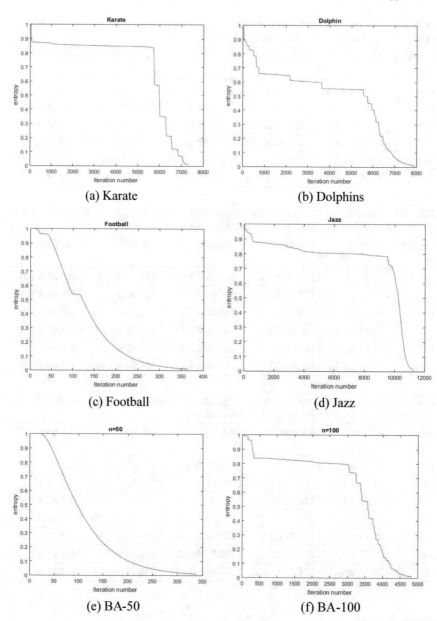

Fig. 3.11 Entropy value of the proposed algorithm on the experimental datasets

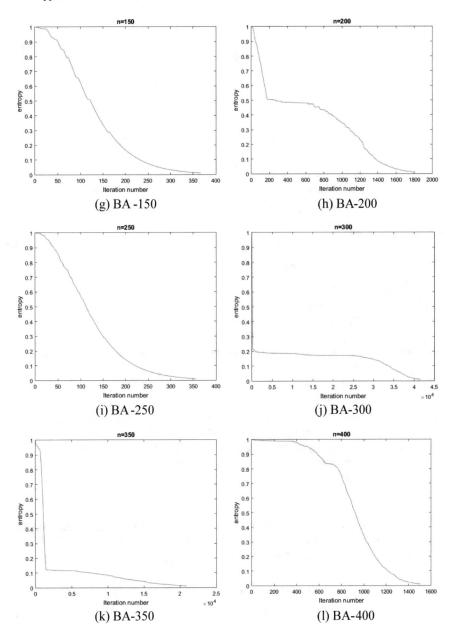

(g) BA -150

(h) BA-200

(i) BA -250

(j) BA-300

(k) BA-350

(l) BA-400

Fig. 3.11 (continued)

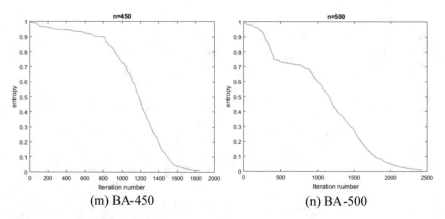

(m) BA-450   (n) BA-500

Fig. 3.11 (continued)

Table 3.7 Description of the algorithms used for experimental comparison

Algorithm	Description
High-degree	A heuristics approach which selects k nodes with maximum degree
HITS	Hyperlink-Induced Topic Search (HITS; also known as hubs and authorities) is a link analysis algorithm that rates Web pages
PageRank	This algorithm is used for ranking web pages according to their importance in a web graph
CELF	A greedy algorithm which executes CELF using a Monte Carlo simulation
CELF++	An improved version of CELF
K-Shell	This algorithm is an update for MDD [1] which considers both the residual degree and exhausted degree, respectively
Pareto-Shell	An improved version of K-Shell algorithm with the Pareto optimal set consisting of spreads, with the ratio of high out-degree to in-degree and high in-degree

For PageRank, we have selected the k highest ranked nodes as initial nodes and we used $\varepsilon = 10^{-4}$ as a stopping condition for convergence of the PageRank algorithm.
– In HITS, the stopping condition was set to 1.02 for selecting the k initial seed nodes.
– CELF was run 10,000 times to evaluate and estimate the spread of the seed set.

The details of the algorithms which is used to provide comparisons are presented in Table 3.7.

In this experiment, different criteria are used such as random node selection, High-degree nodes, High betweenness nodes, High closeness nodes, and High eccentricity nodes as an initial node. The initial activated nodes are supposed to be (5, 10, 15, 20, and 25%) of the whole network. In each case, a specific percentage of the nodes are selected according to one of the assumed criteria as the initial activated nodes. The details are shown in Fig. 3.12 as a activated node for different numbers of the

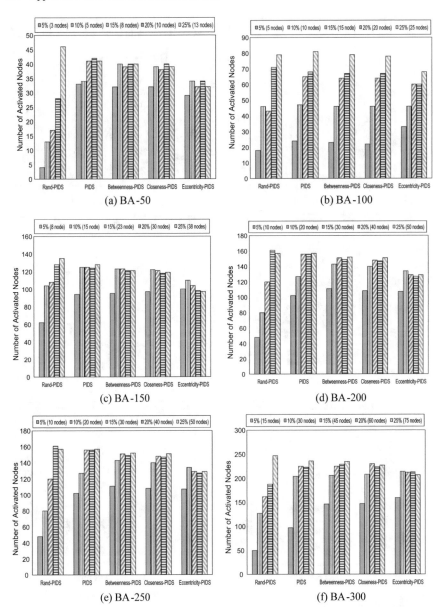

Fig. 3.12 The number of activated nodes using different centrality and initial numbers of activated nodes

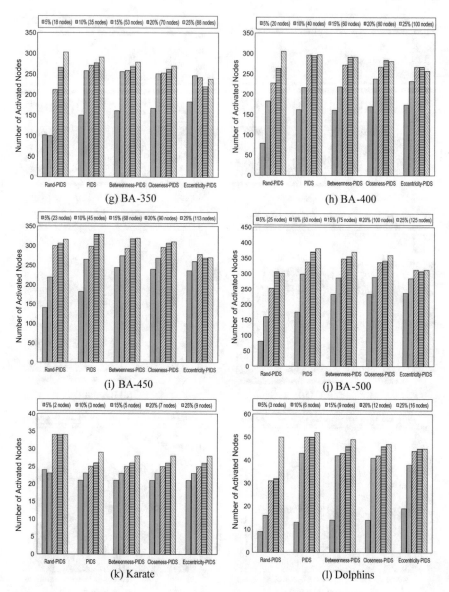

(g) BA-350

(h) BA-400

(i) BA-450

(j) BA-500

(k) Karate

(l) Dolphins

Fig. 3.12 (continued)

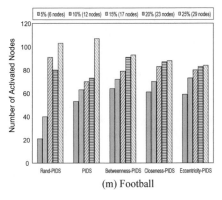
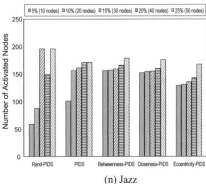

(m) Football   (n) Jazz

Fig. 3.12 (continued)

initial active nodes for the different criteria. In these figures, the short forms of
the criteria are used as follows: Rand-PIDS, PIDS, Betweenness-PIDS, Closeness-
PIDS, and Eccentricity-PIDS stand for random, High degree, High Betweenness,
High Closeness, and High Eccentricity criteria.

In comparison with other algorithms, our proposed algorithm works as similar
to the other heuristics in terms of High-degree PIDS. We note that the term High-
degree PIDS refers to the PIDS in which the nodes are sorted based on their degree. In
some real networks, such as Dolphins and Football, the High-degree PIDS naturally
has a high influence spread, since these networks consist of many group structure
with dense connectivity into intergroup and sparse connectivity with respect to other
part of the network with High-degree nodes; hence, the greatest influence spread
is expected to be obtained by selecting a High-degree node. In addition, in other
datasets such as BA-50, BA-150, BA-250, Karate, and Jazz, the number of activated
nodes at the end of the proposed algorithm is as close as in the other algorithms.

Experiment 4

This experiment was conducted in order to study the performance of the proposed
algorithm (PIDS) in terms of the number of activated nodes. In this experiment, we
intend to compare the proposed RW-PIDS algorithm with other algorithms (Rand,
High-degree [18], PageRank [19], HITS [20], CELF [21], and CELF++ [22]) in terms
of choosing number of seed sets size. Figure 3.13, depicts the number of activated
nodes using different algorithms. The effect of the size of seed sets is also investigated
in this experiment. We assume the sizes of the seed sets to be 5, 10, 15, 20, and 25%
of the whole network.

From the results, we may conclude that by increasing the size of the seed nodes at
the end of all algorithms, the number of activated nodes increases gradually; however,
in some cases, it is clear that the proposed RW-PIDS algorithm outperforms the
other algorithms which mean that *RW-PIDS* nodes in those networks are essential
for diffusion, particularly in synthetic datasets.

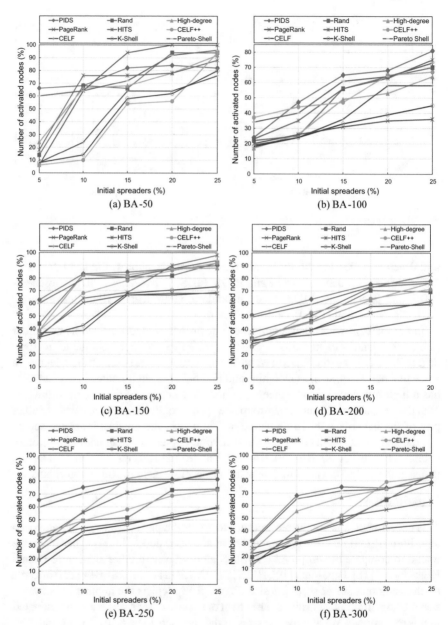

Fig. 3.13 Numbers of activated nodes using different algorithms and initial numbers of activated nodes

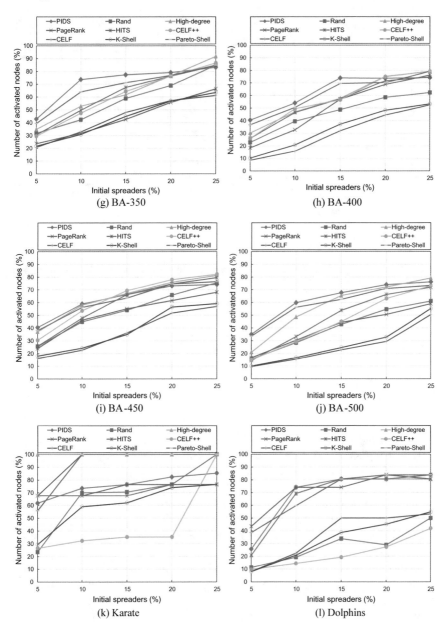

Fig. 3.13 (continued)

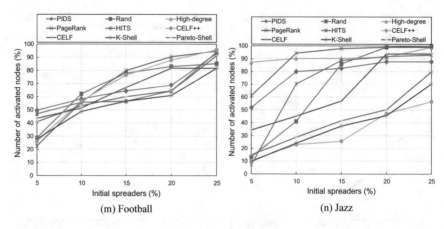

(m) Football   (n) Jazz

Fig. 3.13 (continued)

Table 3.8 Percentage of outperformance for the proposed algorithm (MPIDS) in comparison with other algorithms based on the numbers of activated nodes

	Rand (%)	High-degree (%)	PageRank (%)	HITS (%)	CELF++ (%)	CELF (%)	K-Shell (%)	Pareto-Shell (%)
Synthetic networks	84	70	92	72	78	100	100	82
Real networks	70	10	30	35	80	65	100	70

To show the superiority of the proposed algorithm more clearly, a precise data analysis is carried out below on the numbers of activated nodes. To present an accurate analysis of Fig. 3.13 and to demonstrate the efficiency of our proposed algorithm in producing larger numbers of activated nodes, the percentage of outperformance by the proposed algorithm is given in Table 3.8. To compute the percentage of outperformance, it is sufficient to find the number of test problems in which the proposed algorithm achieves greater numbers of activated nodes than each of the existing algorithms.

As can be seen in Table 3.8, the proposed algorithm achieves higher numbers of activated nodes in a considerable portion of the test problems for synthetic networks, in comparison with the other algorithms. However, for real networks, the proposed algorithm is significantly more efficient than Rand, CELF++, and CELF, and there is no meaningful difference between the performance of the proposed algorithm and that of High-degree, PageRank and HITS. Moreover, our proposed algorithm outperforms the recently proposed algorithms K-Shell and Pareto-Shell in most test problems on real networks.

3.4.3 Conclusion

In this section, an application of *IRW-LA* model for finding minimum positive influence dominating set in social networks is presented. As mentioned above, by the cooperation of the learning automata and using learning mechanism, the promising links are detected and proper result has been obtained for positive influence dominating set. Due to exponential complexity of minimum positive influence dominating set problem, the algorithm is not supposed to reach an optimal solution in reasonable time, but the results show that the proposed algorithm optimizes solutions than other well-known algorithms in minimum positive influence dominating set problem.

References

1. Kwok YK (2011) Peer-to-peer computing: applications, architecture, protocols, and challenges. CRC Press, United States
2. Tschorsch F, Scheuermann B (2016) Bitcoin and beyond: a technical survey on decentralized digital currencies. IEEE Commun Surv Tutorials 18:2084–2123
3. Saghiri AM, Meybodi MR (2016) An approach for designing cognitive engines in cognitive peer-to-peer networks. J Netw Comput Appl 70:17–40. https://doi.org/10.1016/j.jnca.2016.05.012
4. Chawathe Y, Ratnasamy S, Breslau L, Lanham N, Shenker S (2003) Making gnutella-like p2p systems scalable. In: Proceedings of the conference on applications, technologies, architectures, and protocols for computer communications. ACM, Karlsruhe, Germany, pp 407–418
5. Clarke I, Sandberg O, Wiley B, Hong T (2001) Freenet: a distributed anonymous information storage and retrieval system. Designing privacy enhancing technologies. Springer, Berkeley, USA, pp 46–66
6. Stoica I, Morris R, Karger D, Kaashoek MF, Balakrishnan H (2001) Chord: a scalable peer-to-peer lookup service for internet applications. ACM SIGCOMM Comput Commun Rev 31:149–160
7. Ratnasamy S, Francis P, Handley M, Karp R, Shenker S (2001) A scalable content-addressable network. ACM SIGCOMM Comput Commun Rev 31:161–172
8. Gkantsidis C, Mihail M, Saberi A (2006) Random walks in peer-to-peer networks: algorithms and evaluation. Perform Eval 63:241–263
9. Ghorbani M, Meybodi MR, Saghiri AM (2013) A new version of k-random walks algorithm in peer-to-peer networks utilizing learning automata. 5th Conference on information and knowledge technology. IEEE Computer Society, Shiraz, Iran, pp 1–6
10. Ghorbani M, Meybodi MR, Saghiri AM (2013) A novel self-adaptive search algorithm for unstructured peer-to-peer networks utilizing learning automata. 3rd joint conference of AI & Robotics and 5th RoboCup Iran Open International Symposium. IEEE, Qazvin, Iran, pp 1–6
11. Li X, Wu J (2006) Improve searching by reinforcement learning in unstructured P2Ps. In: 26th IEEE international conference on distributed computing systems workshops. IEEE, Portugal
12. IBM Knowledge Center—Social Network graph. https://www.ibm.com/support/knowledgecenter/en/SS2HSB_8.1.0/com.ibm.iis.ii.analyzingvis.doc/topics/eas_ilog_con_socialnetworkgraph.html. Accessed 29 Sept 2018
13. Brin S, Page L (1998) The anatomy of a large-scale hypertextual web search engine. Comput Netw ISDN Syst 30:107–117
14. Kernighan BW, Lin S (1970) An efficient heuristic procedure for partitioning graphs. Bell Syst Tech J 49:291–307

15. Quick review of graph mining with R. In: Quick review of graph mining with R. http://www. lumenai.fr/blog/quick-review-of-graph-mining-with-r. Accessed 29 Sept 2018
16. Wang F, Du H, Camacho E, Xu K, Lee W, Shi Y, Shan S (2011) On positive influence dominating sets in social networks. Theoret Comput Sci 412:265–269
17. Baumgart I, Heep B, Krause S (2007) OverSim: a flexible overlay network simulation framework. IEEE global internet symposium. IEEE Computer Society, Anchorage, USA, pp 79–84
18. Kempe D, Kleinberg J, Tardos É (2003) Maximizing the spread of influence through a social network. In: Proceedings of the ninth ACM SIGKDD international conference on knowledge discovery and data mining. ACM, New York, USA, pp 137–146
19. Brin S, Page L (1998) The anatomy of a large-scale hypertextual Web search engine. Comput Netw ISDN Syst 30:107–117
20. Kleinberg JM (1999) Authoritative sources in a hyperlinked environment. J ACM (JACM) 46:604–632
21. Leskovec J, Krause A, Guestrin C, Faloutsos C, VanBriesen J, Glance N (2007) Cost-effective outbreak detection in networks. In: Proceedings of the 13th ACM SIGKDD international conference on knowledge discovery and data mining. ACM, USA, pp 420–429
22. Goyal A, Lu W, Lakshmanan LV (2011) Celf ++: optimizing the greedy algorithm for influence maximization in social networks. In: Proceedings of the 20th international conference companion on World wide web. ACM, India, pp 47–48

Chapter 4
Conclusions and Future Directions

Abstract In this book, we presented a comprehensive analysis of the principal tasks related to the intelligent models of random walk based on learning automata. After introducing the random walk, we focused on its main drawbacks and weak performance in real-world applications. Then, three intelligent models were established on the bases of random walk. Moreover, theoretical analysis and convergence behavior of the proposed models based on weak convergence theory and Ordinary Differential Equation (ODE) were studied. In addition, the proposed models were applied in two large-scale complex networks such as peer-to-peer networks and social networks. It should be noted that this book presents a new horizon for future research based on random walk and learning systems. The rationale behind the proposed models can be extended with other learning techniques such as Q-learning. In the other hand, there are numerous versions of random walk which should be reconfigured with learning methods in real-world applications. In this chapter, the conclusions and the future directions are explained in more detail.

Keywords Intelligent models of random walk · Learning systems · Q-Learning

4.1 Conclusions

The main conclusions of this book can be summarized as follows:

I. We have presented three intelligent models of random walk based on random walk algorithms and learning automata theory. We note that these three models are domain independent and believe that the proposed models are able to be used in other problems and lead to problem-solving in other domains such as big data computation, wireless sensor networks, and other domains related to complex systems.

II. The first model (*IKRW-LA*) was used to predict k promising link for each node
     in peer-to-peer networks. Based on the policy of the algorithm, a node in the
     network is selected randomly and triggers k-random neighbors of the selected
     node. The process of triggering nodes is continued for all of the triggered nodes.
     In this model, each node utilizes a *KSALA* to trigger the k of the best neighbors
     in order to improve its performance.

III. The second model, which is called Self-Organized Intelligent Random Walk
     based on Learning Automata *(SOIRW-LA)* to resolve and increase the ability of
     the *IKRW-LA* model for path prediction is proposed. In the *k-random* walk algo-
     rithm, determining the optimal value for k is a difficult task and challenging
     problem. Determining a large value for k may generate useless information of
     the network and a small value for k may cause probability of selecting desir-
     able parts of the network decreases. Hence, the *SOIRW-LA* utilizes two actions
     variable-structure learning automaton for each link, and then tries to improve
     the performance of the model.

IV. The third model, which is called *IRW-LA*, is a dynamic graph in which each node
     is equipped with a variable-structure learning automaton and the structure of the
     network changes with time. The advantage of this model than others models
     is to predict the best path of the network considering the feedback received
     from the network. This model is equivalent to the first model (*IKRW-LA*) when
     the value of parameter k is equal to one. To evaluate the performance of this
     model, the entropy criterion has been used to examine the convergence of the
     algorithm.

V.   We have given the theoretical analysis and convergence results of the *IRW-LA*
     model. To achieve this goal, by the use of weak convergence theorems, the
     proposed model is approximated by an *Ordinary Differential Equation (ODE)*
     [1]. Then, we have shown that the resulting ODE converges to the solution of
     an optimization problem.

4.2   Future Directions

This book has made a positive contribution toward using intelligent random walk
as a tool with prediction capabilities in peer-to-peer and social networks. Moreover,
as the notable achievements of this book, it opens up several directions for further
study and research (Fig. 4.1), and some of which are given below.

- Intelligent random walk algorithms based on learning automata

In this book, all of the proposed models of intelligent random walk are designated
based on variable-structure learning automata. One of the research areas that can be
followed in the future is to use different models of learning automata with various
functionalities such as fixed-structure [2], variable-structure [2], distributed [3], and
cellular [4, 5] in the learning mechanism for all the proposed models. Selecting the

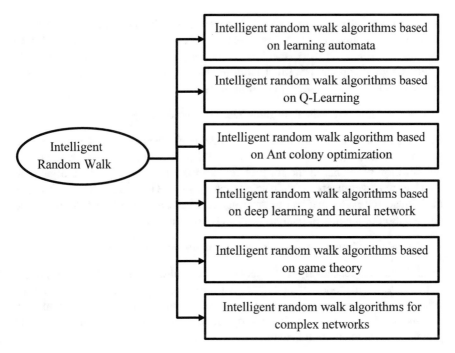

Fig. 4.1 Future directions for the intelligent random walk algorithms

learning algorithms depends on the nature of the problem and can be used in the proposed model for different contexts. These combinations may take the advantage of being comprehensive, distribution, and scalable in the process of prediction which can be used to solve a wide variety of problems.

- Intelligent random walk algorithms based on Q-Learning

All the proposed models of intelligent random walk utilized the learning automata theory to organize their feedback loops. It should be noted that both learning automata and Q-learning theories belong to reinforcement techniques [6]. Therefore, as a future direction, one might redesign our proposed models utilizing Q-learning theory.

- Intelligent random walk algorithm based on ant colony optimization

One direction in the scope of intelligent random walk is to guide random walk with the aid of ants, which take the advantages of both local and global merits for evaporating or depositing pheromones [7]. It seems that the convergence results of this type of random walk can be used to study the convergence of the proposed models given in Sect. 2. In addition, due to the similarity between particle swarm optimization and ant colony, it can be a open similar research area for the future.

- Intelligent random walk algorithms based on deep learning and neural network theories

As mentioned before, all the proposed models of intelligent random walk utilized the learning automata theory to organize their feedback loops. The theories of deep learning and neural networks can be also used to implement the feedback loops in a wide range of applications [8]. Therefore, one may use the mentioned theories to design new models of intelligent random walk.

- Intelligent random walk algorithms based on game theory

Since the intelligent random walk-based model presented in this book works well in solving optimization problems and has the ability to learn, it is possible to choose optimization problems in the form of finding the equilibrium point in a game in which each player meets a certain amount of knowledge in the search space [9]. One of the research paths that can be investigated as research work in this area is the combination of intelligent random walk and game theory.

- Intelligent random walk algorithms for complex networks

In this book, we focused on peer-to-peer networks and social networks as the applications of intelligent models of random walk. Both of these networks belong to complex networks. As future works, one may use the intelligent models of the random walk for other complex systems such as biological networks, collaboration networks, and protein–protein interaction networks.

- Intelligent random walk algorithms for positive influence dominating set in power law graphs

As it was previously mentioned, one of the proposed models of intelligent random walk called *IRW-LA* model, has the capability of predicting the most promising path in the network considering the feedback received from the network and the user-defined metrics. This capability was used to find the positive influence dominating set in the social networks [10]. It seems that this solution can be extended to power law social networks. Moreover, considering the nature of the all proposed models in terms of crawling the networks, we may suggest investigating some other problems such as shortest path, longest path, and dominating set.

- Stochastic positive influence dominating set: an intelligent random walk approach

Due to the unpredictable and dynamic nature of user behavior and human activities in social networks, their structural and behavioral parameters are time-varying parameters and for this reason, using deterministic graphs for modeling the behavior of users may not be appropriate [11]. Stochastic positive influence dominating set refers to a novel problem because of the nature of traverse in the social networks. The notable point is that the entire models of intelligent random walk may be effective to solve the mentioned problem in this scope based on biased sampling strategies. It should be noted that learning automata used by all the proposed models were used as a powerful tool for biased sampling in social graph in [12, 13].

References

1. Walter W (1998) Ordinary differential equations. Springer
2. Narendra KS, Thathachar MAL (1989) Learning automata: an introduction. Prentice Hall
3. Khomami MMD, Rezvanian AR, Meybodi MR (2016) Distributed learning automata-based algorithm for community detection in complex networks. Int J Mod Phys B 30:1650042
4. Beigy H, Meybodi MR (2004) A mathematical framework for cellular learning automata. Adv Complex Syst 3:295–319
5. Rezvanian A, Saghiri AM, Vahidipour M, Esnaashari M, Meybodi MR (2018) Recent advances in learning automata. Springer
6. Sutton RS, Barto AG (1998) Reinforcement learning: an introduction. Cambridge University Press
7. Ma T, Xia Z, Yang F (2017) An ant colony random walk algorithm for overlapping community detection. International conference on intelligent data engineering and automated learning. Springer, China, pp 20–26
8. Haykin S (1994) Neural networks: a comprehensive foundation. Prentice Hall PTR
9. Osborne MJ, Rubinstein A (1994) A course in game theory. MIT press
10. Wang F, Camacho E, Xu K (2009) Positive influence dominating set in online social networks. In: Du D-Z, Hu X, Pardalos PM (eds) Combinatorial optimization and applications. Springer, Berlin Heidelberg, pp 313–321
11. Rezvanian A, Meybodi MR (2016) Stochastic graph as a model for social networks. Comput Hum Behav 64:621–640
12. Rezvanian AR, Meybodi MR (2015) A new learning automata based sampling algorithm for social networks. Int J Commun Syst. https://doi.org/10.1002/dac.3091
13. Rezvanian A, Meybodi MR (2017) A new learning automata-based sampling algorithm for social networks. Int J Commun Syst 30:e3091

Printed in the United States
By Bookmasters